Bookworms

Creepy Crawlers

Worms

Jack K. Patterson

Cavendish Square
New York

Published in 2019 by Cavendish Square Publishing, LLC
243 5th Avenue, Suite 136, New York, NY 10016

First Edition

Library of Congress Cataloging-in-Publication Data

Names: Patterson, Jack K., 1988- author.
Title: Worms / Jack K. Patterson.
Description: New York : Cavendish Square, 2019. | Series: Creepy crawlers | Audience: Grades K-3. | Includes bibliographical references and index.
Identifiers: LCCN 2018026117 (print) | LCCN 2018027044 (ebook) | ISBN 9781502643063 (ebook) |
ISBN 9781502642721 (library bound) | ISBN 9781502643049 (pbk.) | ISBN 9781502643056 (6 pack)
Subjects: LCSH: Worms--Juvenile literature.
Classification: LCC QL386.6 (ebook) | LCC QL386.6 .P38 2019 (print) | DDC 592/.3--dc23
LC record available at https://lccn.loc.gov/2018026117

Editorial Director: David McNamara
Editor: Kristen Susienka
Copy Editor: Nathan Heidelberger
Associate Art Director: Alan Sliwinski
Designer: Megan Metté
Production Coordinator: Karol Szymczuk
Photo Research: J8 Media

The photographs in this book are used by permission and through the courtesy of: Cover Alexander Sviridov/Shutterstock,com; p. 5 D. Kucharski K. Kucharska/Shutterstock.com; p. 7 Heidi & Hans-Juergen Koch/Minden Pictures Getty Images; p. 9 Temmuz Can Arsiray/Getty Images; p. 11 Colombini Medeiros, Fabio/Animals Animals; p. 13 David M. Dennis/Animals Animals; p. 15 Kokhanchikov/Shutterstock.com; p. 17 Lobster20/Shutterstock.com; p. 19 Digoarpi/Shutterstock.com; p. 21 Menna/Shutterstock.com.

Printed in the United States of America

Contents

Worms are creepy crawlers!

They have slimy skin.

They don't have bones, ears, or eyes.

Worms start life inside
a **cocoon**.

A cocoon is like an egg.

It grows a worm inside it.

An adult worm can make
many cocoons.

The cocoons sit in wet dirt underground.

The wet dirt is called soil.

Soon the cocoons **hatch**.

The baby worms leave their cocoons.

They dig into the soil.

Adult worms have long bodies.

They move through the soil.

They help it stay wet.

That is good for the soil.

Worms can be colored brown, pink, or red.

They have more than one heart.

A worm's slimy skin can taste bad to other animals.

This keeps **predators** away.

Worms like to eat **nutrients** in the soil.

Nutrients come from leaves or grass in the soil.

When it rains, worms like to leave the soil.

Have you seen worms on a rainy day?

New Words

cocoon (kuh-KOON) A tight ball that holds and protects baby worms.

hatch (HATCH) When a worm leaves its cocoon.

nutrients (NOO-tree-entz) Healthy ingredients that make up food.

predators (PRED-uh-torz) Animals or insects that like to eat other animals or insects.

Index

About the Author

Jack K. Patterson lives in Boston, Massachusetts. He enjoys gardening, bird watching, and exploring the outdoors. He is an insect enthusiast who has his own bug collection. He enjoys learning about the world around him.

About BOOKWORMS

Bookworms help independent readers gain reading confidence through high-frequency words, simple sentences, and strong picture/text support. Each book explores a concept that helps children relate what they read to the world they live in.